ROCKET SCIENCE FOR KIDS

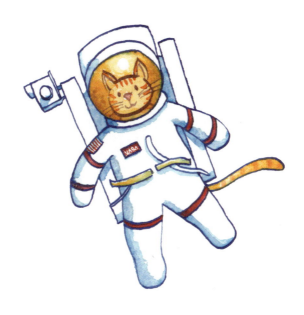

Want more rocket science content for your young learner?

Check out the free book bonuses online for additional educational material you can use at home or in the classroom.

rocketscienceforkids.com/bookbonuses

Books from Hype Hack Publishing are available at special discounts when purchased in quantity for educational use.

For details, contact sales@rocketscienceforkids.com

© 2025 Russell Newman (text) and Laura Fearn (illustrations)

All rights reserved. No part of this book may be reproduced without written permission of the copyright owners, except as permitted by U.S. copyright law. Special thanks to NASA for the public domain satellite photos of the Earth and the Moon.

Published in the U.S.A by Hype Hack Publishing

Printed in the U.S.A

First Edition

ISBN: 979-8-9908136-0-1 (Hardcover)

DISCLAIMER: This is a children's book, not an actual engineering textbook! Many simplifications were made to increase understanding for young audiences. Gosh! The rocket engines don't even have valves or instrumentation depicted. Hype Hack Publishing accepts no liability for exploding rockets, engine hard starts, destroyed ground support equipment, or lack of orbital velocity on your next rocket project. The rocket industry is hard work; please use more resources than just this book when designing a rocket (although this is a great starting point)!

ROCKET SCIENCE FOR KIDS

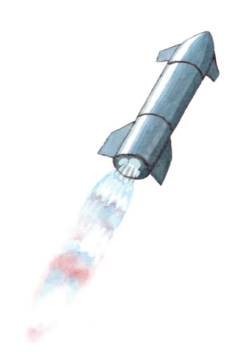

Written by Russell Newman

Illustrated by Laura Fearn

Part 1   What is a Rocket?

Part 2   Basic Rocket Science

Part 3   How Rocket Engines Work

Part 4   How Rocket Tanks Work

Part 5   Rockets on Top of Rockets

Part 6   Want to Build Rockets?

Part 1:
What is a Rocket?

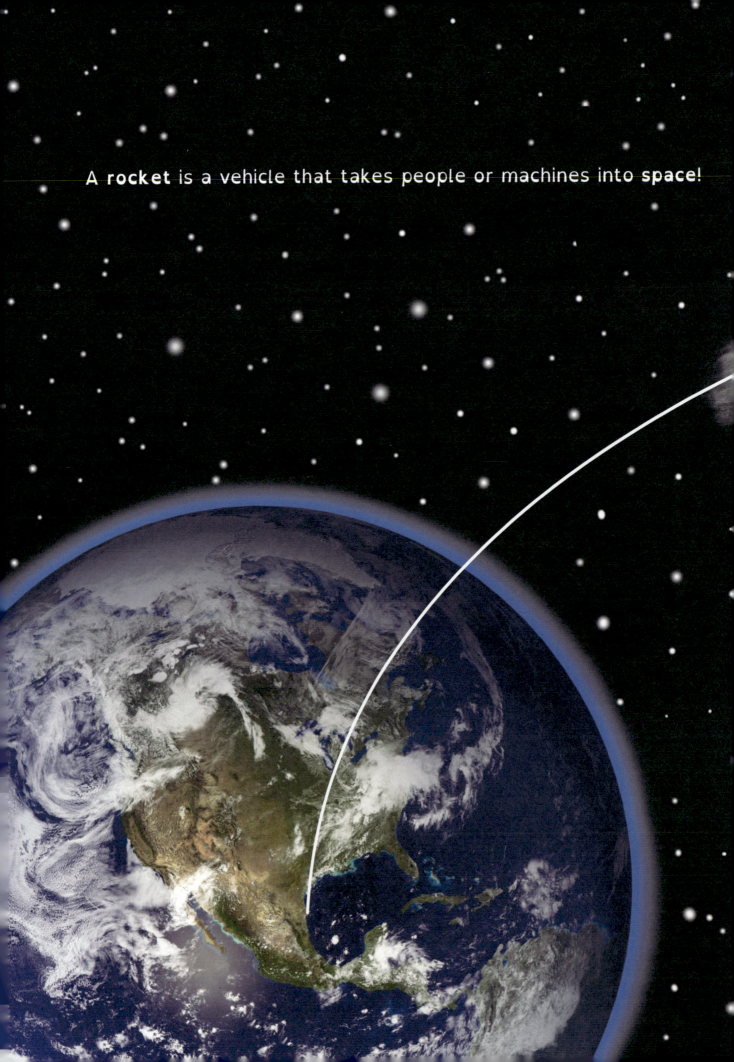

A **rocket** is a vehicle that takes people or machines into **space**!

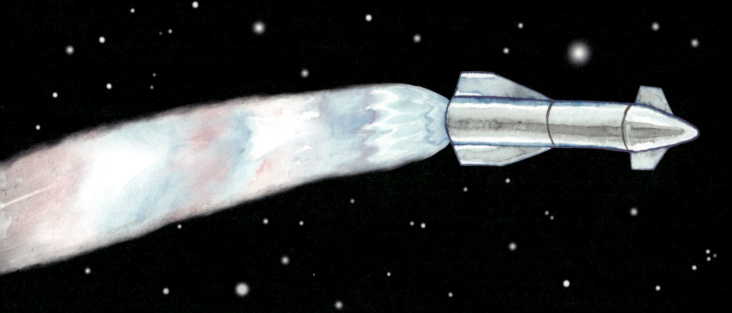

We live on planet Earth.

If you go high enough, there is no more air!

We call this **space**.

Going to **space** can be very useful!

We put machines into space called **satellites**.

Satellites help people navigate even without a map.

People love satellites!

Satellites circle the Earth in an **orbit**.

To orbit, the satellites must be high and going very **fast**!

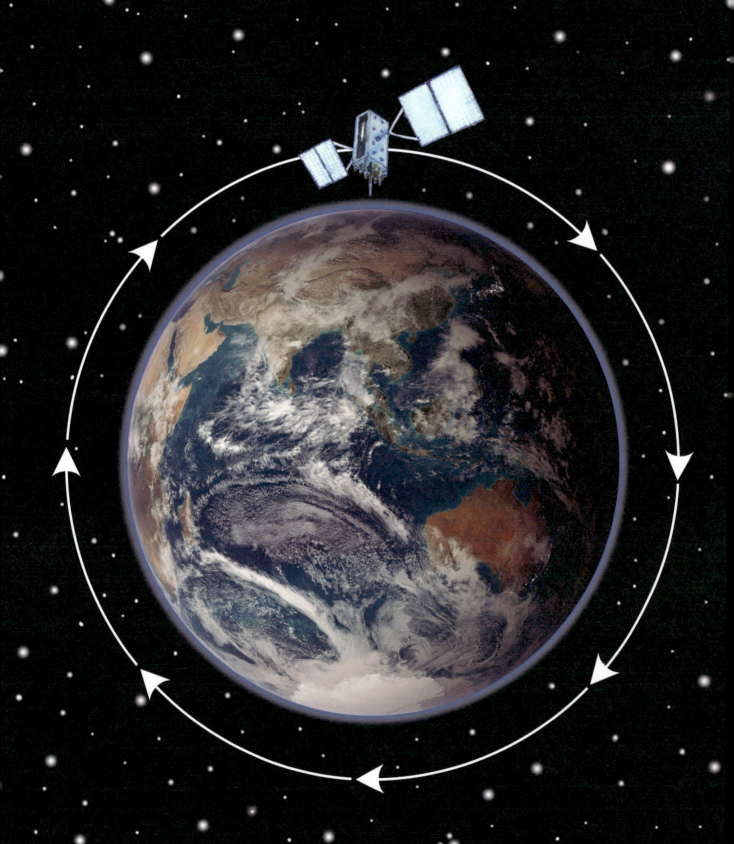

A satellite in orbit is always falling towards the Earth,
but never touches the ground!
It is going so fast that it stays in a circle bigger than the Earth.

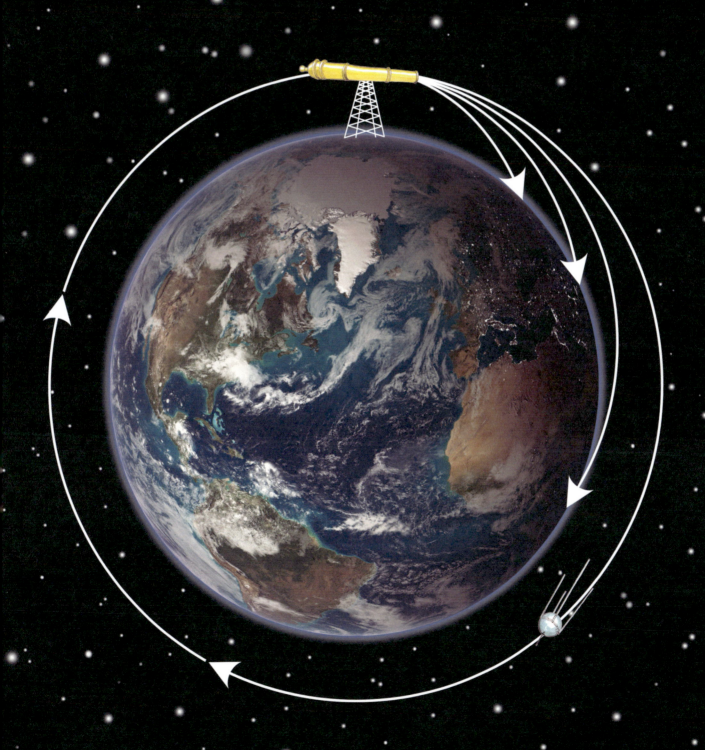

Rockets are how we lift satellites into their orbits.

Making rockets go very, very fast is a challenging problem!

Are you ready to learn the basics of rocket science?

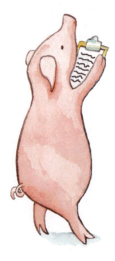

Part 2:
Basic Rocket Science

Rocket science can be simple!

A rocket shoots fire downwards to produce a force that pushes it upwards.

This force makes the rocket go faster

and faster

and faster!

The best rocket can lift the heaviest satellite and push it fast enough to orbit the earth.

Three things determine how fast a rocket can go while carrying a satellite:

1) How fast the fire shoots out of the rocket (faster is better!)

2) How much fuel the rocket can carry (more is better!)

3) How little the rocket weighs (less is better!)

Here's an experiment to try!
If you wear skates and throw a
baseball, you will roll backwards.

The faster you throw,
      the faster you will go!

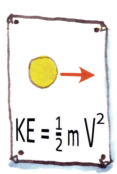

$KE = \tfrac{1}{2} m V^2$

The fuel in a rocket is like millions of baseballs to throw to go faster and faster!

Any weight that is not extra baseballs is weight that makes it harder to go faster and faster.

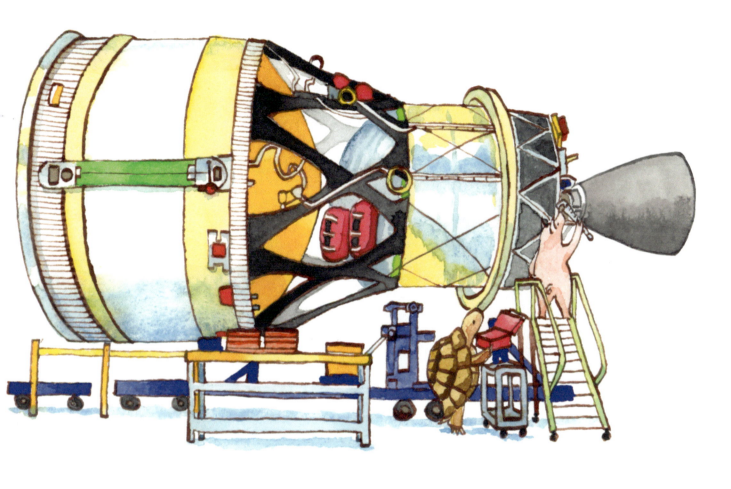

This is why rocket parts are made as lightweight as possible.

Now you know the basics of rocket science:

1) Faster fire!
2) More fuel!
3) Lightweight rocket!

Are you ready to learn how rocket engines work?

Part 3:
How Rocket Engines Work

To push itself, a rocket engine shoots gas backwards in the form of fire.

This fire is made by mixing and burning two liquids very quickly!

These two liquids are **rocket fuel** and **liquid oxygen**.

Rocket fuel is like jet fuel that is used by airplanes.

Liquid oxygen is like the air you breathe, but it is very cold!

So cold that the air turns from a gas into a liquid!

A simple rocket engine takes the **rocket fuel** and **liquid oxygen** from big tanks and pushes them into the engine to burn.

A stronger engine design uses pumps to push the two liquids faster into the engine for burning!

This kind of engine uses a smaller rocket engine to make gas that turns a paddle wheel connected to the two pumps.

This is a good rocket engine design!

Rocket engines with **gas generators** and **turbopumps** are harder to build, but they are much better for lifting satellites to orbit.

Part 4:
How Rocket Tanks Work

When you look at a rocket, most of what you see is tanks!

These tanks store the two liquids that the engines use.

We make tanks lightweight by using pressure!
This is similar to a soda can.

Here's an experiment to try:
You can stand on a fully sealed soda can!

The pressure inside makes it strong, even with very thin walls.

Without the pressure inside, you can crush the can easily.

We pressurize rocket tanks using helium gas.

Helium is the same gas used to fill party balloons!

The rocket stores the helium gas in smaller tanks that sit inside the big tanks.

Pressurized rocket tanks are lightweight.

They are an important part of every good rocket design!

Are you ready to learn about rockets on top of rockets?

Part 5:
Rockets on Top of Rockets

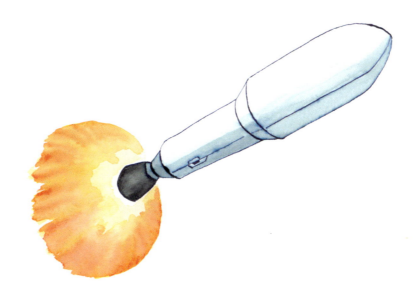

To build a great rocket, you want to stack two rockets on top of each other!

These rockets are called **stages**.

They work together to accomplish the mission.

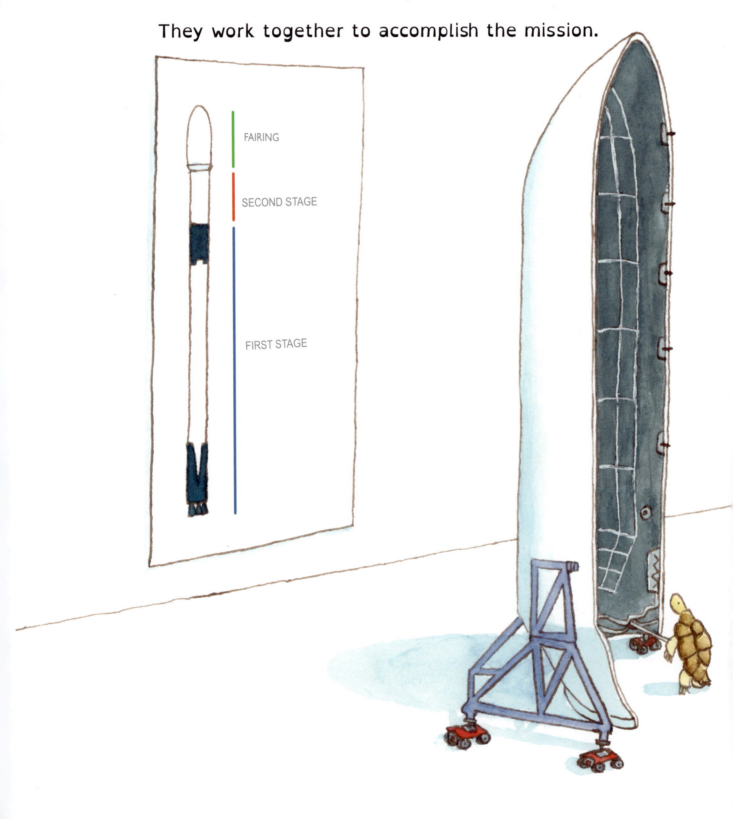

The important satellite sits on top of the whole stack with a protective cover called a **fairing.**

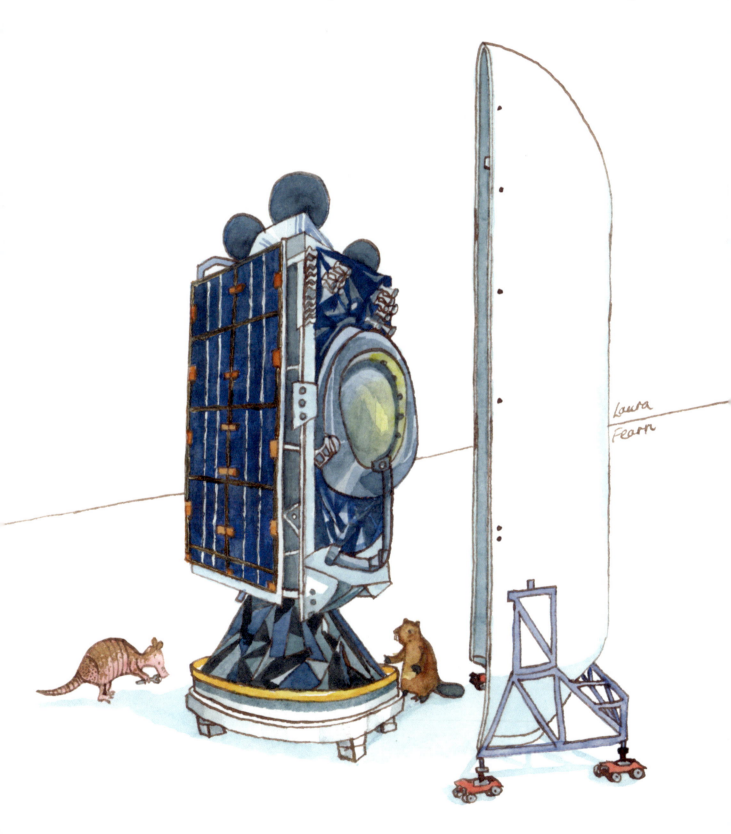

On launch day the rocket is filled with fuel.

Mission Control starts the countdown:

3...

2...

1...

Lift Off!

so that the second stage can start its engines.

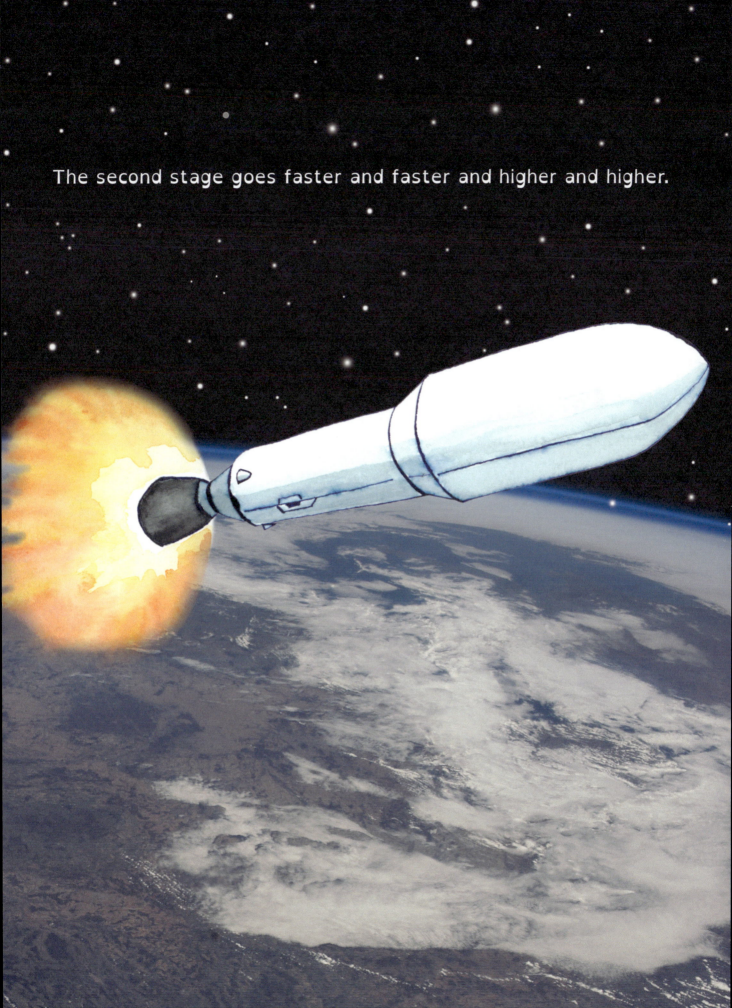
The second stage goes faster and faster and higher and higher.

The first stage's job is done. Now it can go home to fly again!

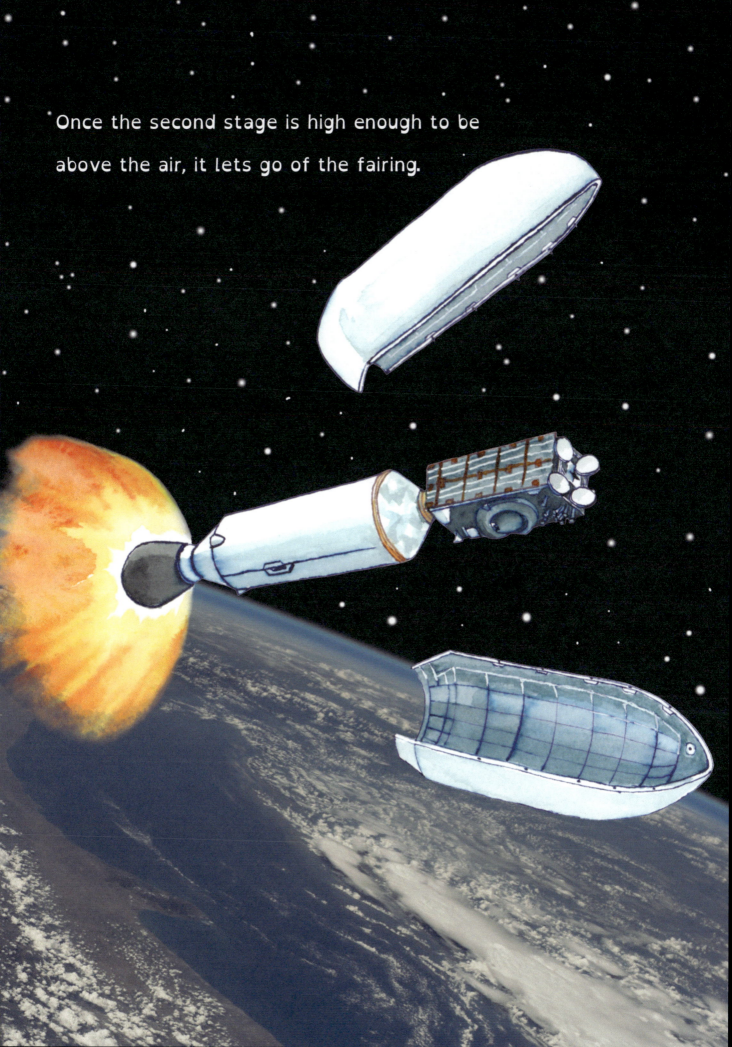

Once the second stage is high enough to be above the air, it lets go of the fairing.

Once the second stage runs out of fuel,

its job is done and it lets go of the satellite.

The satellite is in orbit! Now it will happily circle the Earth helping people all around the world!

Do you want to build rockets?

Part 6:
Want to Build Rockets?

If you want to build rockets, you Should consider becoming an **engineer!**

There are many different types of engineers who design lots of cool things!

Engineers design rockets, cars, airplanes, bridges, buildings, phones, computers, video games, and more!

Engineers work together in teams to solve problems that help people.

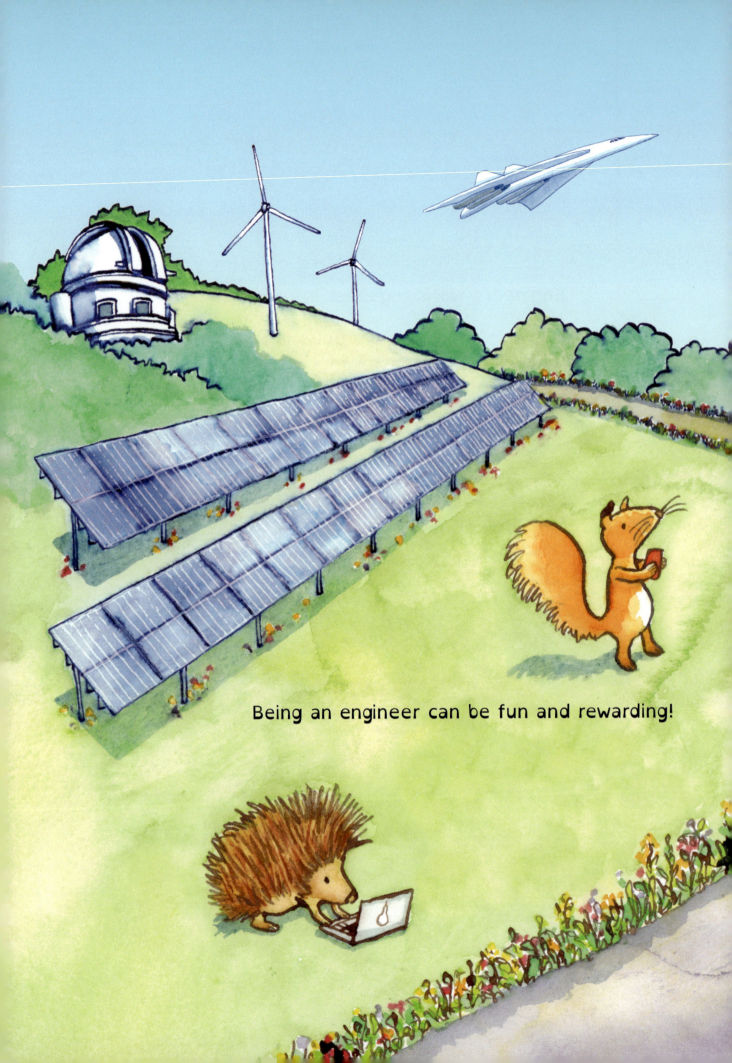
Being an engineer can be fun and rewarding!

Think about being an engineer when you grow up!

The end.

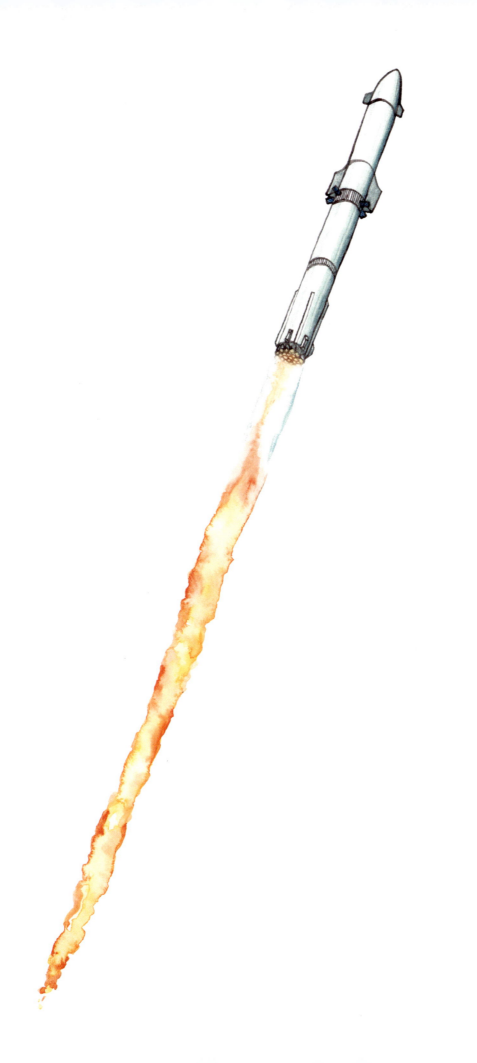

Want more rocket science content for your young learner?

Check out the free book bonuses online for additional educational material you can use at home or in the classroom!

rocketscienceforkids.com/bookbonuses

Books from Hype Hack Publishing are available at special discounts when purchased in quantity for educational use.

For details, contact **sales@rocketscienceforkids.com**

Hey, It's Russ here
I'm the book's author!

I hope you and your kids enjoyed this book!

I have a favor to ask of you. Would you consider giving this book a review on Amazon, please?

Your review on **Amazon,** or other favorite online retailer, will help make this book more discoverable to parents like you. My ambitious goal is to help one million kids learn about engineering! This book can help inspire the next generation of scientists and engineers who will change the world!

Many thanks in advance,

-Russ

amazon.com/review/create-review

Russell Newman

This book was written by a real rocket scientist! Russell is an aerospace engineer who has worked at several rocket companies, including Elon Musk's SpaceX. He holds a degree in Mechanical Engineering from Caltech and is currently working to revolutionize the green energy industry. Russell lives in Los Angeles, California, with his wonderful wife, three kids, and two cats.

Russell can be reached at **rocketscienceforkids.com**

Laura Fearn

Laura is an artist, illustrator, and award-winning designer who paints in acrylic, watercolour, and pen & ink. She studied Illustration at the University of Brighton, UK and worked in London as Senior Designer at a leading Children's Publisher. She works from her garden studio in Bath, UK and lives with her family and pet tortoise. Laura is available for freelance illustration, design work, and custom artwork commissions!

Laura can be reached at **laurafearn.com**